Service Life Evaluation—Design Specification

An ACI Standard

Reported by ACI Committee 365

Kyle D. Stanish, Chair
Jose Pacheco, Secretary

Marwa Abdelrahman	Richard Cantin	O. Burkan Isgor	Karthik H. Obla
James M. Aldred	Larry D. Church	Zoubir Lounis	Bruce G. Smith
Muhammed P. A. Basheer	Carolyn M. Hansson	Matthew A. Miltenberger	Michael D. A. Thomas*
Evan C. Bentz	Doug Hooton	Mohamad Nagi	Wael A. Zatar
Neal S. Berke	Meghdad Hoseini	Paul A. Noyce	Shengjun Zhou

Consulting Members

Antonio J. Aldykiewicz Jr.	Charles D. Pomeroy	Paul G. Tourney	Yash Paul Virmani
David G. Manning	Jesus Rodriguez	Alexander M. Vaysburd	

Liaison Members

Dirk Schlicke
Vute Sirivivatnanon

*Deceased.

This Design Specification provides minimum requirements for performing a service life evaluation as part of the design process for new structures and implementing the results of the evaluation into the construction phase. This Design Specification can be used as part of a design-bid-build project, a design-build project, or other project delivery options. The Design Specification is independent of the specific model or technique used to perform the service life evaluation. Although service life modeling is commonly used to evaluate chloride transport causing corrosion deterioration, the approach outlined in this Design Specification can be used for any deterioration mechanism that is capable of being modeled. The service life engineer performing the evaluation can either be the prime consultant or a subconsultant. A service life report is produced as part of this specification, documenting the service life evaluation, followed by a service life record report documenting the implementation into the new construction.

Keywords: degradation mechanisms; deterministic modeling; durability; performance design; probabilistic modeling; quality control/quality assurance; service life prediction.

CONTENTS

CHAPTER 1—GENERAL, p. 3
1.1—Scope, p. 3
1.2—Purpose, p. 3
1.3—Interpretation, p. 4
1.4—Service life engineer, p. 5

CHAPTER 2—DEFINITIONS, p. 6
2.1—Definitions, p. 6

CHAPTER 3—PROJECT CRITERIA AND DOCUMENTATION, p. 8
3.1—General, p. 8
3.2—Terminology, p. 8
3.3—Project requirements, p. 8
3.4—Basis of design, p. 11

ACI CODE-365-24 was approved by the ACI Standards Board for publication July 2024, and published November 2024.
Copyright © 2024, American Concrete Institute.
All rights reserved including rights of reproduction and use in any form or by any means, including the making of copies by any photo process, or by electronic or mechanical device, printed, written, or oral, or recording for sound or visual reproduction or for use in any knowledge or retrieval system or device, unless permission in writing is obtained from the copyright proprietors.

CHAPTER 4—APPROACH SELECTION, p. 12
4.1—Determination of governing durability limit states, p. 12
4.2—End-of-service criteria, p. 12
4.3—Service life prediction approach selection, p. 12
4.4—Mitigation or avoidance of potential degradation mechanisms, p. 13
4.5—Material performance and construction practices, p. 14
4.6—Accelerated testing, p. 14

CHAPTER 5—MODELING AND INPUT PARAMETERS, p. 15
5.1—General, p. 15
5.2—Model documentation, p. 15
5.3—Input parameters, p. 15
5.4—Deterministic modeling, p. 15
5.5—Probabilistic modeling, p. 15
5.6—Effect of cracks, p. 16

CHAPTER 6—RESULTS AND EVALUATION, p. 17
6.1—General, p. 17
6.2—Documentation, p. 17
6.3—Coordination, p. 17
6.4—Reporting, p. 18
6.5—Periodic updates to project requirements and basis of design, p. 18

CHAPTER 7—IMPLEMENTATION, p. 19
7.1—General, p. 19
7.2—Materials testing, p. 19
7.3—Validation, p. 19
7.4—Repairs during construction, p. 20
7.5—Service life record report, p. 20

COMMENTARY REFERENCES, p. 22
Authored references, p. 22

CODE

CHAPTER 1—GENERAL

1.1—Scope

1.1.1 This Design Specification provides the minimum requirements for performing, implementing, and documenting service life predictions of new concrete structures or portions thereof.

1.1.2 This Design Specification is intended to be adopted by contract for a specific project. The project scope documents shall define the structures or type and number of concrete members or systems for which service life predictions are applicable.

1.1.3 Modifications to this Design Specification can be adopted by contract, but they are not considered part of this Design Specification.

1.2—Purpose

1.2.1 The purpose of this Design Specification is to provide minimum requirements for performing a service life prediction for a new concrete structure.

COMMENTARY

CHAPTER R1—GENERAL

R1.1—Scope

R1.1.1 This Design Specification does not address existing structures. Although the principles for conducting a service life prediction are similar for new and existing structures, performing service life predictions on existing, repaired, or rehabilitated structures can be more challenging. The service life engineer must be aware of current conditions, typically through performing a condition assessment of the structure or members to be repaired before performing service life predictions. Guidance on conducting a condition assessment can be found in ACI 364.1.

R1.1.2 The procedures and requirements outlined in this Design Specification can also be used when not specifically adopted to demonstrate that a comprehensive service life prediction has been performed.

R1.1.3 The American Concrete Institute recommends that this Design Specification be adopted in its entirety. When modifications to the Design Specification are implemented, they become the responsibility of the entity implementing the changes.

R1.2—Purpose

R1.2.1 The procedure outlined in the document is shown in Fig. R.1.2.1.

CODE

1.3—Interpretation

1.3.1 The official version of this Design Specification is the English language version using inch-pound units published by the American Concrete Institute.

1.3.2 In case of conflict between the official version of this Design Specification and other versions of the Design Specification, the official version governs.

COMMENTARY

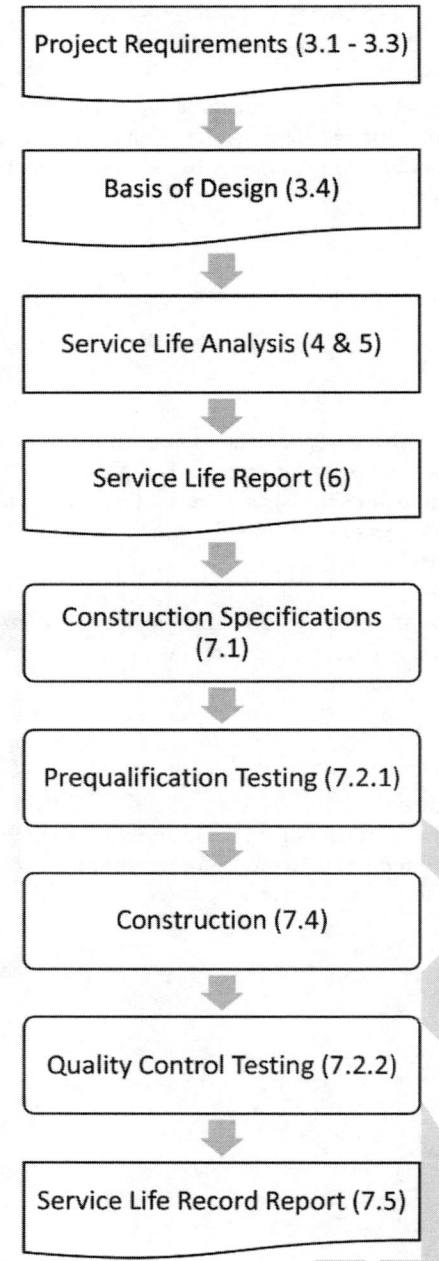

Fig. R1.2.1—Service life prediction process.

R1.3—Interpretation

R1.3.2 If codes or other specifications are adopted that have requirements for service life prediction, the owner is typically responsible for determining which requirement governs. Typically, codes have precedence over Design Specifications.

CODE

1.3.3 This Design Specification includes a commentary. The commentary is intended to provide contextual information and guidance but is not part of the Design Specification, does not provide binding requirements, and shall not be used to create a conflict with or ambiguity in this Design Specification.

1.3.4 This Design Specification shall be interpreted in a manner that avoids conflict between or amongst its provisions. Specific provisions shall govern over general provisions.

1.4—Service life engineer

This Design Specification is prepared for the specialty engineer contracted to perform the service life prediction. For the purposes of this document, the specialty engineer shall be referred to as the service life engineer. The individual(s) with overall project responsibilities shall be referred to as the licensed design professional.

COMMENTARY

R1.3.4 General provisions are broad statements.

CODE

CHAPTER 2—DEFINITIONS

2.1—Definitions

The following terms are defined for general use in this Design Specification.

as-built service life—the estimated service life based on the in-place properties of the structure, including the measured materials properties and as-built concrete cover, and considering documented, intended maintenance.

basis of design—project service life prediction approach as outlined by the service life engineer and accepted by the owner and licensed design professional; includes design service life and definitions, among other project parameters.

contractor—entity responsible for constructing the structure; includes general contractor, their subcontractors such as the concrete contractor, and their vendors such as the materials supplier.

design service life—period after construction of a new structure, member, or assembly during which the durability performance is designed to satisfy the specified durability requirements if maintained as proposed but without being subjected to an overload or extreme event.

deterministic modeling—design based on characteristic input parameter values to provide a single output value.

durability limit states—definitions that correspond to durability-related aspects or performance at a defined point in time.

end-of-service criteria—conditions that correspond to durability-related aspects beyond which specified service requirements for a structure or system are no longer met.

licensed design professional—individual who is licensed to practice engineering as defined by the statutory requirements of the professional licensing laws of the state or jurisdiction in which the project is to be constructed, and who is in responsible charge of the overall design effort.

mitigation of degradation—actions required by service life engineer to reduce the rate or impact of degradation on concrete members or systems.

operator—entity with legal possession and control over property, including the structure.

owner—entity who has legal possession and control of the structure, who may or may not be responsible for operating it after the project is completed.

predicted service life—service life of a structure estimated by means of service life modeling or analysis, considering the documented assumptions.

prequalification testing—materials testing on concrete trial batches prior to the construction of a structure to confirm the proposed concrete mixture meets required properties.

probabilistic modeling—design based on consideration of input parameter values described by statistical distributions. It is typically interpreted by evaluating the output at a certain level of reliability.

project requirements—specific design service life requirements defined by the service life engineer based on project information provided by licensed design profes-

COMMENTARY

CODE

sional, contractor or the owner that define the minimum design service life, the acceptable condition of structure at end of service, the anticipated maintenance schedule and activities during service (if applicable), and any other considerations pertaining to the service life prediction.

service life engineer—individual who is licensed to practice engineering as defined by the statutory requirements of the professional licensing laws of the state or jurisdiction in which the structure is to be constructed, and who is responsible for the service life prediction. The individual shall be qualified by training and experience to perform service life prediction.

service life record report—project record document prepared by the service life engineer outlining the service-life-related aspects of the project. The service life record report must include any and all of the following as applicable: the service life report, qualification requirements, qualification testing results, production records, batch tickets, test reports conducted during construction, as-built concrete conditions, as-built concrete cover thicknesses, and documentation of repairs conducted during construction.

service life report—document prepared by the service life engineer documenting the service life prediction meeting the requirements of this Design Specification.

COMMENTARY

CODE

CHAPTER 3—PROJECT CRITERIA AND DOCUMENTATION

3.1—General

3.1.1 The service life engineer shall define the project service life requirements based on project information provided by the licensed design professional, contractor, or owner.

3.1.2 A basis-of-design document shall be developed by the service life engineer based on the project requirements and accepted by the owner and licensed design professional, if applicable. Service life design deliverables shall be enumerated in the basis-of-design document agreed upon with the owner, the licensed design professional, contractor, and the service life engineer in accordance with the project requirements.

3.2—Terminology

Terminology related to specific requirements of the project and other terms relevant to the specific structure or method(s) selected for service life prediction shall be defined by the service life engineer. These may include but are not limited to: architect/engineer; contractor; service life engineer; licensed design professional; prescriptive requirement; service life prediction model; and other terms relevant to the specific structure to be constructed, and method(s) selected for service life prediction.

3.3—Project requirements

Project requirements shall include at a minimum: 1) design service life; 2) end-of-service-life criteria; 3) design service life reliability requirements or risk considerations; and 4) anticipated maintenance to be conducted during design service life. Project requirements shall consider the importance of the structure or structural member(s) and the consequences of failure to meet durability requirements.

COMMENTARY

CHAPTER R3—PROJECT CRITERIA AND DOCUMENTATION

R3.1—General

R3.1.1 Contractual obligations and expectations should be established at the start of the project. The service life engineer can either be a specialty engineer with responsibility for service life prediction or the licensed design professional for the entire project. The responsibility for meeting the project requirements will contractually funnel to the service life engineer responsible for performing service life prediction. It is assumed that the owner will establish a desired service life for the project and will also provide the anticipated maintenance plan to achieve the desired service life. The service life report should clearly indicate the initial project information for the basis-of-design.

R3.1.2 At times, the owner may provide a detailed basis-of-design document that the service life engineer is expected to adopt. Should this basis-of-design be adopted, the service life engineer remains responsible for the requirements and analyses outlined in the service life report. Establishing the project goals, acceptance criteria, and terminology relevant to service life prediction remains the responsibility of the service life engineer.

Confusion can arise when a clear definition of project scope, tasks, and responsibilities is not provided in the project terminology. This confusion may affect the success of the project or diminish opportunities to improve durability in a given environment. In such cases, the design service life might not be realized.

R3.2—Terminology

Concise, succinctly phrased terminology will serve to provide a baseline understanding of key durability and service life prediction concepts, and document-specific terms related to the structure under consideration. This facilitates clear delineation of roles and responsibilities for all project team members, and the owner's understanding of service life prediction tasks. Definitions for relevant terms are provided in Chapter 2. Definitions for specific terminology remain the responsibility of the service life engineer.

R3.3—Project requirements

Project requirements that determine the design service life should be defined by the service life engineer with input from the owner and licensed design professional. Preparation of a formal project requirements document allows the service life engineer to develop the basis of design and identify any risks or conflicts with the project requirements. Key topics include environmental exposure parameters, concrete materials sourcing and availability, reinforcing material availability, sustainability metrics, anticipated weather conditions, schedule and sequencing of construction operations, site constraints, and anticipated deterioration mechanisms within the design service life. As-built service life

CODE

3.3.1 Identify anticipated deterioration mechanisms and appropriate approaches for achieving the design service life for the structural member(s), component(s), or systems subject to design service life requirements as specified by the service life report. Categorize procedures as one of the following: a) mitigate deterioration mechanisms through material selection, construction practices, or design detailing; b) use prescriptive, deem-to-satisfy approach; or c) design to resist deterioration for a set period as a performance limit state. Consider synergistic opportunities among various approaches with respect to design service life, schedule or sequencing opportunities, mitigating risk, and constructability.

3.3.2 Define the design service life for structural members, systems and components, and overall structure. The design service life shall be defined in years. The design service life may be different for different components.

COMMENTARY

is also dependent upon construction quality assurance/quality control activities, design changes during construction, construction practices and tolerances, and performing routine maintenance.

R3.3.1 Deterioration mechanisms are described in ACI PRC-201.2, and more detailed guidance is available from other documents, such as ACI 222.3 (corrosion), and ACI 224.1 (cracking). Different members within a structure can require different approaches based on accessibility and cost to repair post-construction. Depending on the owner's risk tolerance and the service life engineer design, some structural members may require a longer service life than other members. For example, nonreplaceable components such as foundations require excavation for repair and rehabilitation and could require a 100+ year design service life; bridge deck and substructure systems could require 75-year design service life (AASHTO LFRDUS-8); and replaceable components, such as barriers or bearings, may require only 30-year design service life.

Design to resist deterioration can apply either a deem-to-satisfy approach, a deterministic approach, a probabilistic approach, or a combination of methods and may include the use of safety factors.

R3.3.2 Establishing the design service life is the first step in a service life design process. The target service life can vary from temporary structures or short repair cycles to monumental structures or long periods between repairs. The design service life for a project should consider use, location, importance, and the consequences of deterioration. The design service life is defined at the start of a project and is expressed as a measurable or quantifiable amount. Tolerance and or expected variability in service life predictions should be established.

The type, function, and importance of the overall structure needs to be considered when selecting an appropriate design service life. For example, monumental buildings such as cathedrals may need to be designed to last hundreds of years, whereas an industrial mine with 25 years of ore deposits may only need to be designed to meet the life of the mineral deposit. Additionally, structural members or components are likely to be replaced over the life of the structure without resulting in the end of the structure service life. For example, a bridge deck or parapet walls may be designed to be replaced during the life of the bridge. The service life engineer is advised to verify the minimum service life requirements of the applicable codes (for example, AASHTO LFRD). Structural members or components with design service life different from that of the overall structure should be acceptable to the owner and documented in the basis of design.

The in-service environmental conditions should be considered when establishing the design service life. In severe exposure conditions, long design service lives may not be technically or economically reasonable.

CODE

3.3.3 For each design service life requirement, provide definition of the condition of the member, system, or structure that will be deemed as the end of design service life.

COMMENTARY

The owner should identify the desired design service life for a project considering use, location, importance, and the consequence of deterioration. Table R3.3.2 offers some guidance in this regard, Similar tables are located in EN 1990 (2002), ISO 16204 (2012), and Smith (2001).

Table R3.3.2—Example design service life for various project types (based on Smith [2001])

	Local	Interstate	International	Monumental
Infrastructure	50 to 75 years	75 to 100 years	100 to 150 years	150 to 250 years
Buildings	Low-rise	High-rise	Government, institutional	Monumental
	50 to 75 years	75 to 100 years	100 to 175 years	175 to 500 years
Other	Based on commercial or project needs, 25 to 100 years			

R3.3.3 In a well-planned design with service life prediction, the design service life and the end-of-life criteria must be clearly defined by the service life engineer and understood and accepted by the owner. These criteria are usually selected to correspond to a given durability limit state for the structure. The service life engineer should define realistic design service life goals in a manner that can be quantified, such as "time to the initiation of corrosion," "time to first repair," "time to major rehabilitation," or "time to replacement." For probabilistic analyses, a required reliability should be included. Durability limit states need to be defined for each deterioration mechanism.

In the case of reinforcement corrosion, the design service life is often expressed as a function of the time to corrosion initiation, time to corrosion initiation plus propagation time, or the time to major repairs rather than as a time to critical section loss of the steel reinforcement, as the latter option is difficult to verify. Information regarding modeling approaches can be found in ACI 365.1. Unless otherwise requested by the owner, the use of specific methodologies is at the discretion of the service life engineer. The selection, use, and interpretation of service life prediction tools remain the responsibility of the service life engineer.

Pavement performance is often defined by the extent of transverse slab cracking, transverse joint faulting, transverse joint spalling, and pavement smoothness (Hoerner and Darter 2000). The end of predicted service life is reached when these limit states exceed permissible levels.

Deterioration mechanisms for which no verified service life prediction methodologies are available should be mitigated. Expansive reactions caused by aggregates should be mitigated by the use of material screening or selection. ASTM C1778 provides guidance for mitigating the risk of deleterious aggregate reactions in concrete. When chemical degradation from the concrete surface is involved, the design service life can be calculated as a function of the time to reach a certain degradation depth expressed as a function of the total dimension or the concrete cover. Residual concrete cover is typically needed at the end of the design service life to allow for structural performance.

CODE

3.3.4 Define design service life reliability requirements or risk considerations, when applicable, for either the entire structure or individual components.

3.3.5 Define the type and frequency of maintenance that is anticipated to be performed with regard to service life parameters for every design service life scenario under consideration.

3.3.6 Establish the specific duration and conditions for which the analysis is to be considered reliable within the prescribed probability limits or is to be performed for alternative costs comparisons.

3.4—Basis of design

Incorporate service life prediction project requirements into the basis-of-design document. At a minimum, the project basis-of-design document shall include the type of structure and its design service life, apportioned by members if necessary; proposed service life prediction methods for every project phase; testing requirements for concrete; durability limit states; performance verification methods; anticipated maintenance activities over design service life; and reporting requirements. Structural members, components, or systems that shall meet a design service life must be explicitly stated in contract documents and basis of design.

COMMENTARY

The durability limit states are not fixed values and can be defined by the service life engineer, in standard specifications for multiple structures, or on a project-specific basis.

R3.3.4 Consider the importance of the structure when establishing the level of confidence that the structure or specific system will reach its design service life or, conversely, the risk that it will not.

In addition, the owner's tolerance for risk and the consequences of failure need to be explored and considered in the service life prediction approach. Predicted service life may be calculated with either a deterministic model or a probabilistic model. If the probabilistic approach is used, the specifications should call for a reliability requirement to satisfy a given criterion. The reliability requirement must be established as a function of the importance of the structure and the consequences of failure. When failure of a structural member poses a life safety risk, the probability of failure should be set at an acceptable level prescribed by the relevant building code for ultimate limit states.

R3.3.5 During its service life, repair or rehabilitation work may be necessary on the structure and the owner should account for this. The maintenance requirements should be defined in consultation with the owner, including facility maintenance personnel framing durability limits in this manner allows the realistic application of probability limits in the higher end of the possible interval and design parameters that are more practical in terms of concrete properties and required cover. The service life engineer can then assess the risk associated with a structure or individual member failing to meet its design service life requirements.

R3.3.6 Service life modeling is often an input into life cycle cost-benefit analyses. If the purpose for calculating predicted service life is to assess the relative costs of options in present value, the cost of all anticipated maintenance, repairs, salvage, or residual value of each option at the service life design horizon, need to be included. This service life design horizon should be clearly defined along with the discount rate in any request for professional services. The life cycle cost analysis objective is to identify options that minimize the total cost of ownership, not necessarily the cost of construction.

R3.4—Basis of design

Service life assessment must be integrated in the project schedule from the time it is decided to include it in the specifications. Depending on the method used, time should be allowed for the determination of calculation parameters or the qualification of concrete mixtures and materials

CODE

CHAPTER 4—APPROACH SELECTION

4.1—Determination of governing durability limit states

4.1.1 Identify the anticipated durability limit states. A durability limit state shall consist of: 1) the definition of the corresponding performance level; 2) an evaluation process or test method; and 3) an acceptance criterion. In cases where several durability limit states requirements are applicable, the service life engineer shall determine which durability limit state requirement will govern the design service life of the member, component, or system.

4.1.2 Identify each exposure condition associated with each durability limit state affecting each concrete member or component(s). In cases of combined exposures or conflicting durability limit state requirements, the service life engineer shall define which exposure condition(s) shall govern based on the defined governing durability limit state(s).

4.1.3 The service life engineer shall define the applicable durability limit states for concrete members, components, or system.

4.2—End-of-service criteria

The service life engineer shall define end-of-service criteria for each of the applicable durability limit states. The applicable criteria shall be incorporated into project documentation. End-of-service criteria correspond to the condition of the structure upon which the design service life is deemed to be finished as determined by performance conditions or contract requirements.

4.3—Service life prediction approach selection

4.3.1 The service life engineer shall identify and select the appropriate methodology to model or estimate the predicted service life of concrete subject to the exposure conditions and durability limit states defined in the basis of design.

4.3.2 The selected methodology shall define the procedure for conducting a service life prediction, the necessary input parameters, and the limitations of the methodology as indicated in Chapter 6.

COMMENTARY

CHAPTER R4—APPROACH SELECTION

R4.1—Determination of governing durability limit states

R4.1.1 Durability limit states should be related to serviceability limit states. This means that failure to maintain a specified performance over a defined period can cause economical and operational losses but not cause severe structural damage or collapse. Durability limit states should only be applicable to serviceable conditions. Ultimate limit state conditions should not be used for establishing durability limits.

R4.1.2 Examples of exposure conditions associated with typical durability requirements include exposure to abrasion, aggressive chemicals, freezing-and-thawing, carbon dioxide or monoxide, chloride, elevated temperature, sulfates exposure, and moisture promoting alkali-aggregate reaction. Further information can be found in ACI PRC-201.2. Refined classifications of exposure conditions may be required, such as differentiating between the submerged, tidal, splash, and atmospheric zones in marine environments.

The service life engineer is required to define the exposure condition(s) for each concrete member that requires it. Combinations of exposure conditions often exist. For example, parking garages in cold climates are subject to freezing-and-thawing cycles and chloride exposure from deicing salts.

R4.1.3 Durability limit states can be general and cover multiple members, components, or systems with similar exposure conditions, durability performance, and accessibility for repair.

R4.3—Service life prediction approach selection

R4.3.1 The selected methodology applies to the governing deterioration mechanism and intended concrete performance. When multiple mechanisms are considered, a single methodology or multiple methodologies can be used. ACI 365.1 has examples of service life prediction approaches for different deterioration mechanisms.

R4.3.2 Service life prediction methodologies may consist of concepts, equations, or analyses that can range from historical performance with specific materials to prediction of mechanical property deterioration with time to numerical modeling of ionic transport in concrete. The most widely

CODE	COMMENTARY
	used methodologies can describe the ingress of chlorides by the following approaches: (a) Diffusion through the concrete pore structure—This can range from Cranks' solution to Fick's Second Law to modified diffusion equations with correction parameters for long-term behavior of cementitious systems (b) Transport of multi-ionic species—This type of model considers the transport of other ions into and through the concrete medium (c) Reactive transport models—Novel models describing the transport and reaction of ions in the concrete pore structure could be used for specific conditions. Information regarding modeling approaches can be found in ACI 365.1
4.3.3 The service life engineer shall define the selected methodology and incorporate it in the basis of design.	**R4.3.3** Once the methodology is selected, the service life engineer should make sure that the owner, licensed design professional, contractor, and other involved parties understand the key aspects of the selected methodology and can understand its implementation in the basis of design.
4.3.4 When the selected service life prediction methodology requires accelerated testing, it shall meet the requirements of Section 4.6. If the service life prediction methodology requires numerical modeling, it shall meet the requirements of Chapter 5.	
4.3.5 The service life engineer shall identify if other project requirements could affect the service life prediction, durability limit state, and methodology.	**R4.3.5** Construction often requires that specific concrete members or systems meet multiple specification requirements, including design service life and durability performance. The licensed design professional, in consultation with the service life engineer, should determine the governing specification requirements in case of conflict. Mass concrete or lightweight concrete are examples of other project requirements that could affect service life prediction. Costs should also be a consideration for selecting an approach.
4.4—Mitigation or avoidance of potential degradation mechanisms **4.4.1** It is permissible to define degradation as mitigated or avoided to satisfy an end-of-service criterion.	**R4.4—Mitigation or avoidance of potential degradation mechanisms** **R4.4.1** Mitigation or avoidance of degradation mechanisms with a deem-to-satisfy approach may be selected by the service life engineer for deterioration mechanisms that are complex and for which limited availability of modeling methods for prognosis exist. Typical degradation mechanisms that may be considered mitigated or avoided within the design service life include freezing-and-thawing deterioration and alkali-aggregate reaction.
4.4.2 For durability limit states for which avoidance or mitigation approaches are used, the required material properties shall be specified.	**R4.4.2** ACI PRC-201.2 discusses various durability limit states, including potential approaches to avoid or mitigate deterioration, if applicable.

CODE

4.5—Material performance and construction practices

4.5.1 The service life engineer shall establish the material performance and anticipated construction practices to provide satisfactory compliance with the design service life of components/systems, or other contract documents.

4.5.2 The selection of acceptable material and construction practices shall be consistent with the governing durability limit states, deterioration mechanisms, and economic considerations necessary to achieve the specified design service life requirements.

4.6—Accelerated testing

4.6.1 Accelerated testing using more severe exposure conditions can be used to establish the predicted service life of the concrete structure.

4.6.2 The acceleration factor shall be determined prior to performing the accelerated testing, and its basis shall be documented.

COMMENTARY

R4.5—Material performance and construction practices

R4.5.1 Projects in which the design and material performance requirements for achieving the specified design service life are known from past experience can fall into this category. Examples are the use of nonshrink grout, overlays, or other protective systems.

R4.6—Accelerated testing

R4.6.1 During an accelerated testing program, a more severe environment is used to increase the rate of degradation. This can involve an increase in concentration of reactants, temperature, humidity, hydraulic pressure, or electrical potential. Use of accelerated testing is discussed further in ACI 365.1.

R4.6.2 As an example, if the degradation is assumed to proceed at a proportional rate by the same mechanism in both accelerated test and in-service, the predicted service life may be estimated from the testing results using Eq. (R4.6.2), where L_{pred} is the predicted service life, L_{test} is the test duration, and k is the acceleration factor.

$$L_{pred} = kL_{test} \quad (R4.6.2)$$

The methodology for determining the acceleration factor should be documented in the basis of design report.

CODE

CHAPTER 5—MODELING AND INPUT PARAMETERS

5.1—General

Modeling procedure and input parameters used shall be documented with sufficient detail for the analysis to be evaluated and reproduced.

5.2—Model documentation

Models used for service life prediction shall have sufficient documentation of the service life prediction concept used and validation of implementation to allow replication of the results.

5.2.1 The concept used to perform the modeling shall be documented in sufficient detail to allow a third party to evaluate the applicability of the general modeling approach.

5.2.2 Implementation of the model approach shall be confirmed by validation with documented experimental results from laboratory or field testing.

5.3—Input parameters

5.3.1 For each input parameter, document the value used in the service life analysis and its basis.

5.3.2 The variability of the input value shall be considered when establishing the values that can be used. This can either be explicit when a probabilistic model is used or implicit when a deterministic model is used.

5.4—Deterministic modeling

Input parameters shall be established with an adequate factor of safety to limit the risk of exceeding values that would reduce the performance of the structure and result in not meeting the required design service life objective. The input parameters definition shall consider the variability of conditions and the consequence(s) of failure.

5.5—Probabilistic modeling

5.5.1 For a probabilistic modeling approach, the variability of all significant input parameters and the correlation between these parameters shall be considered.

COMMENTARY

CHAPTER R5—MODELING AND INPUT PARAMETERS

R5.1—General

Service life modeling can be performed in multiple ways and with multiple objectives, and there is not one standard model or methodology that can encompass all conditions or scenarios. As such, it is necessary to document the approach and assumptions being made to allow review by other practitioners.

R5.2—Model documentation

R5.2.1 Model documentation can be accomplished by reference to relevant published papers or by other documentation of the approach being used. Information regarding modeling approaches can be found in ACI 365.1.

R5.2.2 It is recommended to include by reference a comparison between results obtained with the service life model and observations in the laboratory or from existing structures to verify implementation of the model for the deterioration mechanisms as applicable in the project.

R5.3—Input parameters

R5.3.1 Input parameters can be sourced from values provided by literature, local test data of similar structures, local material properties, testing, public information, or values provided by owner, as appropriate for each specific project, with justification for selection of the source and the value.

R5.4—Deterministic modeling

Deterministic models provide a theoretical basis for examining the relative importance of various factors that influence the performance of a structure or component. For example, the model can assist in making informed decisions regarding a concrete mixture design, cover, reinforcement type, protection, and other design decisions.

R5.5—Probabilistic modeling

R5.5.1 The impact of the correlation between different parameters can be considered using methods such as Monte Carlo analysis, the first order reliability method, or through simplified procedures such as those reported in Bentz (2003). Other stochastic methodologies that could be used

CODE

5.5.2 Deterministic values can be used for parameters that are sufficiently well established.

5.5.3 Document the reliability index that is being adopted for the durability limit states being considered.

5.6—Effect of cracks

5.6.1 The service life engineer shall define if cracking of concrete is considered in the selected service life prediction methodology. If cracking is not considered, follow 5.6.2. If cracking is considered, follow 5.6.3.

5.6.2 If the effect of cracks is not being considered, the service life engineer shall define the acceptable levels of cracking that render the selected service life prediction methodology to still be valid.

5.6.3 Document the approach to incorporate the impact of cracking on the service life prediction. Define the level of cracking that has been incorporated in the service life prediction.

5.6.4 Define the applicable mitigation efforts if cracking exceeds the thresholds defined in 5.6.2 or 5.6.3. The cracks exceeding these limits shall be repaired in accordance with Chapter 7.

COMMENTARY

include Markov chains, Bayesian theory, or Latin Hypercube Sampling.

R5.5.2 Typical parameters that are assigned deterministic values in a probabilistic analysis include background chloride content.

R5.5.3 For durability limit states such as the onset of corrosion, a reliability index of 1.3 is typical (*fib* 2006). In comparison to ultimate limit states in structural design such as collapse, a higher reliability index must be considered as determined by the local structure code but is not needed for durability limit states.

Alternatively, the owner and service life engineer may choose to establish a different risk of failure if not restricted by code requirements.

R5.6—Effect of cracks

R5.6.1 There are multiple methods to account for the effect of cracks on service life predictions. The service life engineer should select the approach that best fits the selected service life prediction methodology and provide predictions for cracked and uncracked conditions.

Because concrete will not be crack-free, the service life engineer should define the types of cracks that are anticipated (controlled) and those that are to be avoided (uncontrolled). An example of controlled cracks is drying shrinkage or restrained cracking. Uncontrolled cracks can be caused by plastic shrinkage, plastic settlement, or others.

R5.6.2 When the effect of cracks is neglected, acceptable levels of cracking should be explicitly stated. An example of reasonable crack widths is included in ACI 224. For example, crack orientation, crack widths, crack spacing, crack depth, or crack density can be used to define the durability limit state conditions upon which cracks are not to be considered in the service life prediction. The service life engineer shall provide guidance on corrective measures should these exceed the defined acceptable level.

R5.6.3 There are many approaches to incorporating cracking in service life prediction models for chloride-induced corrosion, although there is not an industry consensus for a recommended method.

CODE	COMMENTARY
CHAPTER 6—RESULTS AND EVALUATION	**CHAPTER R6—RESULTS AND EVALUATION**

CODE

CHAPTER 6—RESULTS AND EVALUATION

6.1—General
The results shall be reviewed by the service life engineer to verify conformance with the requirements for the project.

6.2—Documentation
6.2.1 The service life engineer shall have a copy of all documents pertinent to the evaluation of the predicted service life of structural components, prequalification testing reports to demonstrate compliance with specification requirements, and information related to the construction practices that may affect the structure predicted service life.

6.3—Coordination
6.3.1 The different members of a structure shall be clearly defined, and their service life performance shall be coordinated so that all the members of the structure achieve the design service life.

6.3.2 The target performance for each of the members shall reflect the accessibility of the member for future inspection and monitoring, as well as repair and replacement.

COMMENTARY

CHAPTER R6—RESULTS AND EVALUATION

R6.2—Documentation
R6.2.1 The service life engineer and licensed design professional should have all documents available for ensuring that the design parameters, concrete material performance, and construction practices have been conducted in accordance with contract documents.

R6.3—Coordination
R6.3.1 Different members of a structure, even made with the same material, can have different performance due to different combinations of exposure, micro-environment, materials composition, design, and construction practices.

A typical example would be different microclimates within a structure depending on whether the location is in or above the splash zone and its exposed surface orientation. If this structure, such a column, extends over different exposure zones (for example, submerged, tidal, splash, or atmosphere zones) and has different face orientations, the difference in its performance should be assessed individually. If different designs for different locations or face orientations are not practical for one structure, the worst-case scenario may be considered in the design to achieve a more conservative structure assessment. However, if different designs are practical, it can be more economical to adjust design parameters (such as cover, additional coating, or alternative reinforcement) for different members to provide uniform performance.

Another example would be a structure that is made of different components that in some cases may include different concrete mixture designs or ingredient materials or constructed using different methods. For example, a bridge may have precast driven piles, cast-in-place pile caps, piers and bent caps, precast beams, and cast-in-place deck slabs. In this case, each component can be assessed individually to achieve the uniform design service life requirement. Consideration should be given to provide increased resistance for similar components exposed to more severe conditions. For example, a pier cap located under an expansion joint may be exposed to more chloride and moisture than intermediate pier caps. This condition may justify special treatment for pier caps located under expansion joints.

R6.3.2 It is common for foundations, for example, to be expected to have a long design service life without repairs whereas the relatively accessible concrete deck on a bridge can be allowed to go through multiple rehabilitation cycles through the design service life of the bridge. The bridge piers, bent caps, beams, and, sometimes, pile caps can be also accessible for monitoring and repair but not for

CODE

6.3.3 Compare predicted service life to project requirements. Service life predictions shall be evaluated to ensure that the proposed design is appropriate.

6.4—Reporting
The service life engineer shall prepare a service life report documenting project requirements; reference codes; service life prediction approach, including assumptions or limitations; input parameters; assumed maintenance; and predicted service life for each of the concrete members and durability limit states considered.

6.5—Periodic updates to project requirements and basis of design
Revise the basis of design as the project progresses if new information or requirements affects the basis of design. If the basis of design revisions requires changes to the project requirements, provide the necessary changes in writing to the owner and licensed design professional.

COMMENTARY

replacement. However, the repair on these members is often limited to small, localized areas. These differences need to be considered during design to achieve uniform service life performance.

R6.3.3 Service life prediction results should meet the design service life; however, excessive conservatism may be undesirable. Overly conservative results may lead to an increase in cost and other issues with the concrete mixture properties such as heat gain, increased shrinkage, and reduced workability.

R6.5—Periodic updates to project requirements and basis of design
Project requirements and basis of design documents should be considered "living" documents and subject to continuous revision, with owner acceptance, as the design and construction process proceeds. It is through these documents that the owner's verification of quality can be managed, and ultimately the life-cycle management of the new construction achieved. Documentation requirements for life-cycle management of concrete structures is defined in ISO 22040.

CODE

CHAPTER 7—IMPLEMENTATION

7.1—General

The service life engineer shall incorporate the information collected in the basis of design, testing results, and other pertinent information into applicable technical specifications, drawings, or construction requirements.

7.2—Materials testing

The service life engineer shall develop materials testing requirements based on the results of the service life evaluation to confirm that the as-built components of the constructed structure meet the material requirements to obtain the design service life.

7.2.1 *Prequalification testing*—The service life engineer shall establish prequalification testing requirements for trial batches of each concrete mixture to be used in the structure. Testing requirements for fresh and hardened concrete shall include test method, including any modifications to standard test procedures; test age; curing requirements; sample requirements; and acceptance criteria, including tolerance.

7.2.2 *Quality control testing*—The licensed design professional and the service life engineer shall develop the requirements for quality control testing during concrete placement. Testing requirements for fresh and hardened concrete shall include test method, including any modifications to standard test procedures; test age; test frequency and number; sample requirements; and acceptance criteria, including tolerances.

7.2.2.1 *Results*—All test results shall be submitted to the service life engineer. Test results that do not meet the acceptance criteria shall be evaluated by the service life engineer and the licensed design professional to identify appropriate remedial measures.

7.3—Validation

7.3.1 Validation of as-built service life shall be conducted at the direction of the service life engineer to the extent required by the owner and defined in the project documents. A conformity evaluation of the completed work shall

COMMENTARY

CHAPTER R7—IMPLEMENTATION

R7.1—General

To build structures that meet the design service life requirements, the results of the service life evaluation need to be incorporated into requirements that can be specified for the contractor to implement.

In addition to the properties that relate to the deterioration mechanisms that were the subject of the service life evaluation, other material durability requirements are needed. Appropriate material properties to prevent other deterioration mechanisms, such as alkali-aggregate reaction and resistance to freezing and thawing, need to be specified as discussed in 4.4.

R7.2—Materials testing

Frequently, owners are placing more responsibility for quality control on the contractor. This practice is justified as long as the owner also maintains a quality assurance presence to ensure that the contractor's quality control plan is being implemented as documented.

Quality control test results should demonstrate compliance with the modeled performance parameters and need to incorporate the impact of normal production variability on the net performance of the completed structure. Establishing test acceptance values needs to be done by the service life engineer because the service life evaluation used may require specialized tests be conducted using the project-specific materials to validate the evaluation.

R7.2.2 *Quality control testing*—Quality control testing can either be the test procedures that were performed during the prequalification testing, or on suitable correlated short-term tests that have had acceptance values established during the prequalification testing at the discretion of the service life engineer. For example, while the prequalification testing may require ASTM C1556, which requires a minimum of 63 days before results can be available, quality control testing may use ASTM C1202, ASTM C1876, or chloride migration coefficient (NTBuild 492), which can be completed in less time. A correlation between different test methods should be established in the prequalification stage.

The service life engineer may consider the variability of the testing results as outlined in ASTM D6607.

R7.2.2.1 *Results*—Remedial measures are discussed in 7.4.2. If the deficiency is not significant, then the owner may choose to accept the construction as is, possibly with appropriate contractual remedies. Proposed remedial measures should be acceptable to the owner.

R7.3—Validation

R7.3.1 The intent of the validation is to demonstrate that the as-built structural members are expected to achieve the specified design service life within the specified reliability targets. This typically includes quality control reports, mate-

CODE

be specified and conducted, and the results documented to be used as a basis for validation. The project specification shall specify requirements for the record documentation of the project. The documentation shall include as-built input parameters used in service life evaluation such as diffusion coefficient and cover thickness. The validation shall demonstrate that the as-built structural members are expected to achieve the design service life within the specified reliability targets.

7.3.2 Should the predicted service life of the members considered not meet the design service life, corrective actions shall be proposed to achieve the specification requirements.

7.3.3 Once validation and any remedial actions have been completed, the service life engineer shall report to the owner, contractor, and other parties involved as required by project requirements on the validation of the predicted service life compliance.

7.3.4 If corrective actions are necessary to achieve the design service life, the validation report shall describe the procedures, materials, and construction practices.

7.4—Repairs during construction
In the event of a nonconformity, repairs necessary to meet the assumptions used in the service life modeling shall be developed by the service life engineer and the licensed design professional.

7.4.1 *Cracking*—Maximum crack size and crack density shall be specified as outlined in 5.6. Repair requirements shall be provided if cracking is not in compliance with allowable cracking requirements.

7.4.2 *Placement or construction defects*—Repair requirements for placement or construction defects shall be specified.

7.4.3 Quality control test results that do not meet the requirements shall be evaluated by the service life engineer. Remedial measures, if necessary, shall be developed to meet the design service life requirements in conjunction with the licensed design professional. Validation must include impact of remedial measures that have been implemented.

7.5—Service life record report
7.5.1 The service life engineer shall prepare a service life record report.

COMMENTARY

rial testing results, preplacement concrete inspection documentation of cover, and concrete cover survey of representative areas. The testing and collection of the as-built material properties is often the responsibility of the contractor, and this should be included in the project specifications. If provisions are not made to collect this information, the service life engineer will be unable to perform the validation.

R7.3.2 When concrete members are not expected to achieve the specification requirements, corrective actions such as the use of overlays, surface protective systems, repair, or replacement of the structural member may need to be considered.

R7.4—Repairs during construction
Repairs during construction may be necessary for other reasons, such as structural requirements or aesthetic requirements. These repairs should meet the design service life.

Necessary repairs can be communicated to the contractor by provision of performance requirements that the repaired member should meet, or by provision of repair procedures.

R7.4.1 *Cracking*—Crack mapping can be useful in evaluation of compliance with cracking criteria as well as design of repairs.

R7.4.2 *Placement or construction defects*—Placement or construction defects include honeycombing, misplaced reinforcing steel, and improper compaction. Repair procedures can either be specified in advance or developed on a case-by-case basis.

R7.4.3 Remedial measures may include additional testing, evaluation of actual conditions, application of coatings or sealers, demolition and replacement of affected members, or similar items.

R7.5—Service life record report
R7.5.1 Portions, such as gathering of records, may be delegated to the contractor, provided the service life engineer reviews the collected data.

CODE

7.5.2 The service life record report shall include a representative sample of the following:
(a) The soil and groundwater testing
(b) Annual temperature and relative humidity cycles
(c) Exposure conditions
(d) Concrete and aggregate testing
(e) Cement and supplementary cementitious material certificates
(f) Reinforcement mill sheets
(g) Service life report
(h) Prequalification testing records
(i) Quality control testing
(j) Construction photographs
(k) Record drawings showing as-built concrete cover
(l) Any protective coatings that were used, as applicable

This can be achieved by reference to other documents that have been prepared and are to be preserved. Verify that the materials used meet or exceed the modeling assumptions.

7.5.3 Deficient areas shall be identified for special maintenance or mitigation practices.

7.5.4 Submit service life record report to owner to be maintained for the life of the structure.

COMMENTARY

R7.5.3 Special maintenance or mitigation practices include treatments such as demolishment of defective work and replacement, crack injection, or application of penetrating sealers or coatings. Photographs shall be taken of any deficient areas before and after repairs are carried out. The deficient areas shall be quantified.

R7.5.4 The service life record report should be considered a "living" document and updated with the results of future work. This document should be maintained as a record document and be provided to design professionals when they do future work on the structure.

The service life record report may include future monitoring recommendations to verify the structure performance based on the actual in-service materials and exposure conditions.

COMMENTARY

COMMENTARY REFERENCES

American Association of State Highway Transportation Officials (AASHTO)
LFRDUS-8-2017—LFRD Bridge Design Specifications

American Concrete Institute (ACI)
ACI PRC-201.2-23—Durable Concrete—Guide
ACI 222.3-11—Guide to Design and Construction Practices to Mitigate Corrosion of Reinforcement in Concrete Structures
ACI 224-01(08)—Control of Cracking in Concrete Structures
ACI 224.1-07—Causes, Evaluation, and Repair of Cracks in Concrete Structures
ACI 364.1-19—Guide for Assessment of Concrete Structures before Rehabilitation
ACI 365.1-17—Report on Service Life Prediction

ASTM International
ASTM C1202-22—Standard Practice for the Electrical Indication of Concrete's Ability to Resist Chloride Ion Penetration
ASTM C1556-22—Standard Test Method for Determining Apparent Diffusion Coefficient of Cementitious Mixtures by Bulk Diffusion
ASTM C1778-23—Standard Guide for Reducing the Risk of Deleterious Alkali-Aggregate Reaction in Concrete
ASTM C1876-23—Standard Test Method for Buk Electrical Resistivity or Bulk Conductivity of Concrete
ASTM D6607-21—Standard Practice for Inclusion of Precision Statement Variation in Specification Limits

European Standards
EN 1990:2002 – Eurocode – Basis of Structural Design

International Standards Organization (ISO)
ISO 16204:2012 – Service Life Design of Concrete Structures
ISO 22040:2021—Lifecycle Management of Concrete Structures

Nordtest
NTBuild 492-99 - Chloride Migration Coefficient form Non-Steady State Migration Experiments

Authored references
Bentz, E. C., 2003, "Probabilistic Modeling of Service Life for Structures Subjected to Chlorides," *ACI Materials Journal*, V. 100, No. 5, Sept.-Oct., pp. 39-47.

fib, 2006, "Model Code for Service Life Design," *Bulletin 34*, Fédération Internationale du Béton, Lausanne, Switzerland, 116 pp.

Hoerner, T. E., and Darter, M. I., 2000, "Improved Prediction Models for PCC Pavement Performance-Related Specifications," *Volume II: PaveSpec 3.0 User's Guide*, FHWA-RD-00-131

Smith, B. G., 2001, "Durability of Silica Fume Concrete Exposed to Chloride in Hot Climates," *Journal of Materials in Civil Engineering*, V. 13, No. 1, Jan-Feb, pp. 41-48.